THIS COLORING BOOK

★ BELONGS TO: ★

☆ MY: ☆

_____

_____

GAVE ME THIS

COLORING BOOK

Funny math
Coloring book

This funny book will help your child to get basic mathematics skills in easy and funny way - they get acquainted with numbers, simple calculations, geometric shapes, etc.
Book contains of 30 pages of interesting exercises with easy instructions, which make study easy and interesting.

Try - your child will be glad to engage in it with you.

Grade: Preschool
Ages: 3 - 5

ISBN-13: 978-1717307880

ISBN-10: 717307884

Author: Allie W. Gartman

Illustrator: Olga Lytvynova

HI! LET US LEARN A NUMBER 1.
Colorize, count and write proper number in the white window!

1

Now we are learning a number 2.
Colorize, count and write proper number in the white window!

2

Now we are learning a number 3.
Colorize, count and write proper number in the white window!

3

Now we are learning a number 4.
Colorize, count and write proper number in the white window!

Now we are learning a number 5.
Colorize, count and write proper number in the white window!

Now we are learning a number 6.
Colorize, count and write proper number in the white window!

Now we are learning a number 7.
Colorize, count and write proper number in the white window!

Now we are learning a number 8.
Colorize, count and write proper number in the white window!

8

Now we are learning a number 9.
Colorize, count and write proper number in the white window!

Geometric shapes

Draw something that looks like geometric shapes below.

Circle

Oval

Triangle

Square

Geometric shapes

Draw something that looks like geometric shapes below.

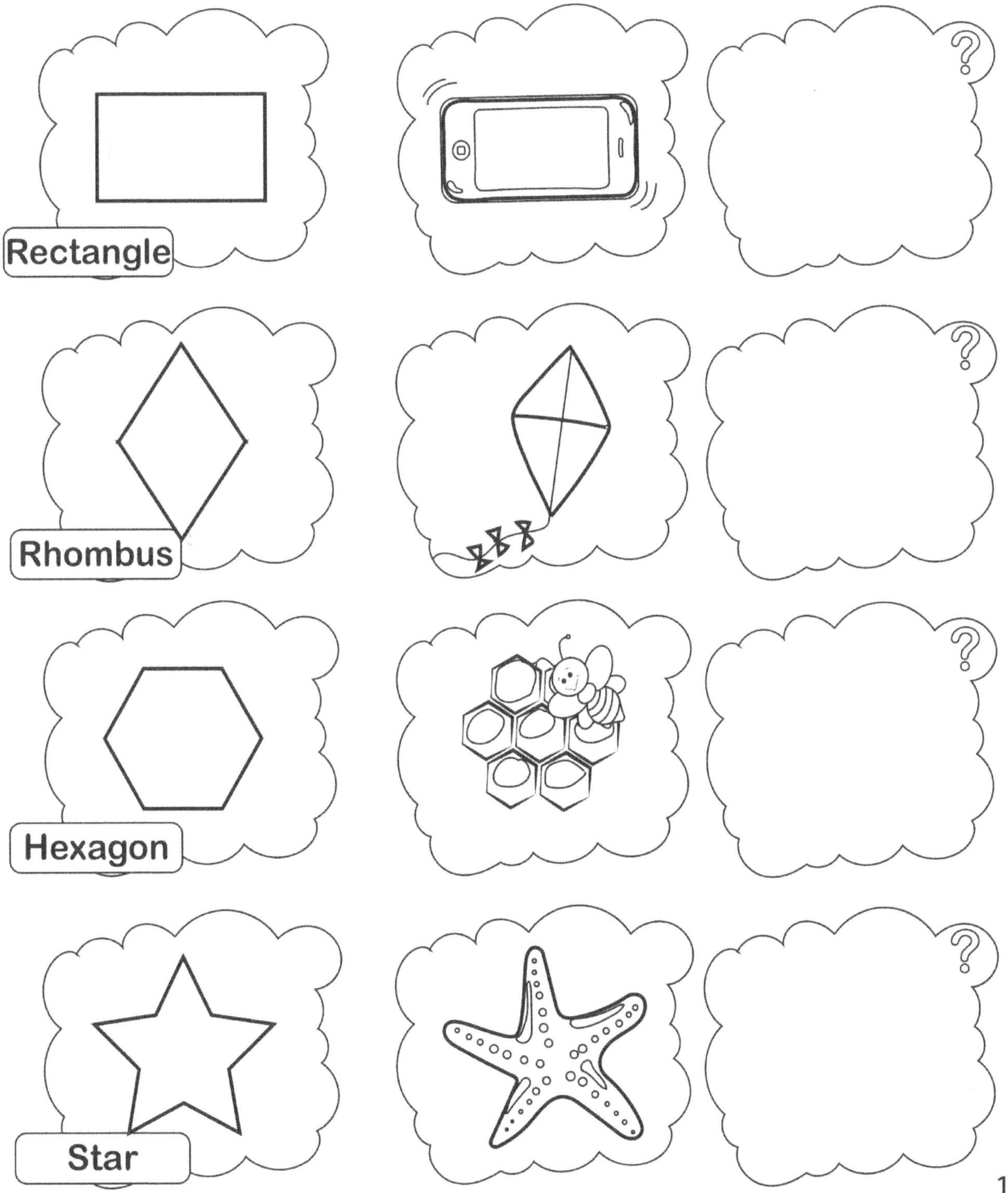

Rectangle

Rhombus

Hexagon

Star

12

Geometric shapes

Help a small giraffe to color and count geometric shapes.

Geometric shapes

Where the geometric shapes are hidden?

Oval

EVEN AND ODD NUMBERS.

An odd number is any number that cannot be divided by 2.
An even number is any number that can be divided by 2.
Count the candies at the picture.
- Colorize the white window to red color if the number of candies is even.
- Colorize the white window to green color, if the number of chocolates is odd.

15

EVEN AND ODD NUMBERS.

Count the candies at the picture.
- Colorize the white window to red color if the number of candies is even.
- Colorize the white window to green color, if the number of chocolates is odd.

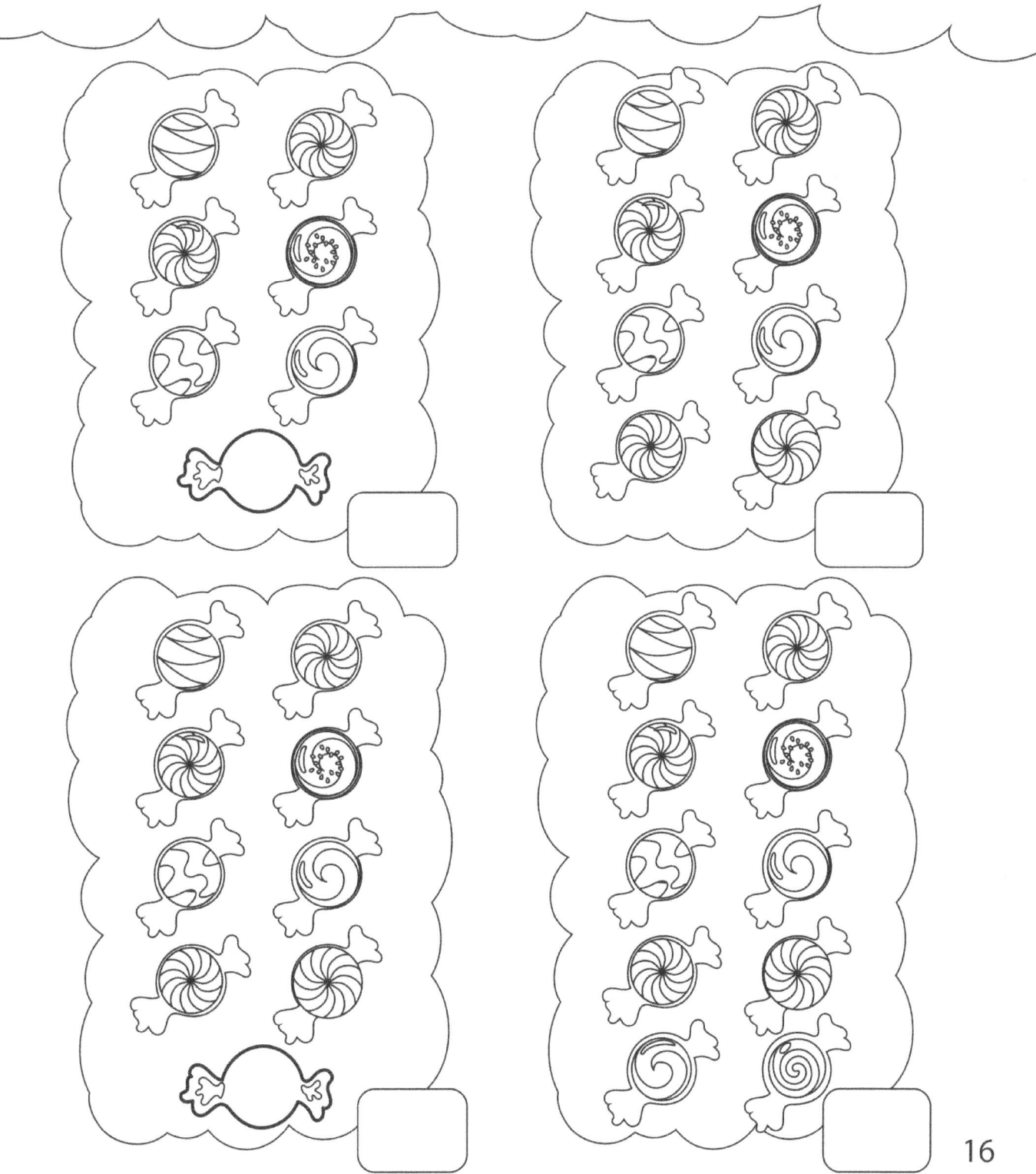

16

Colorize, count and
write proper figure in the white window!

The sun ☐ Bear ☐ Tiger ☐
Tree ☐ Giraffe ☐ Birdie ☐

17

How many?

How many?

19

How many?

3

Addition

How many turtles?
Write proper figure in the white window!

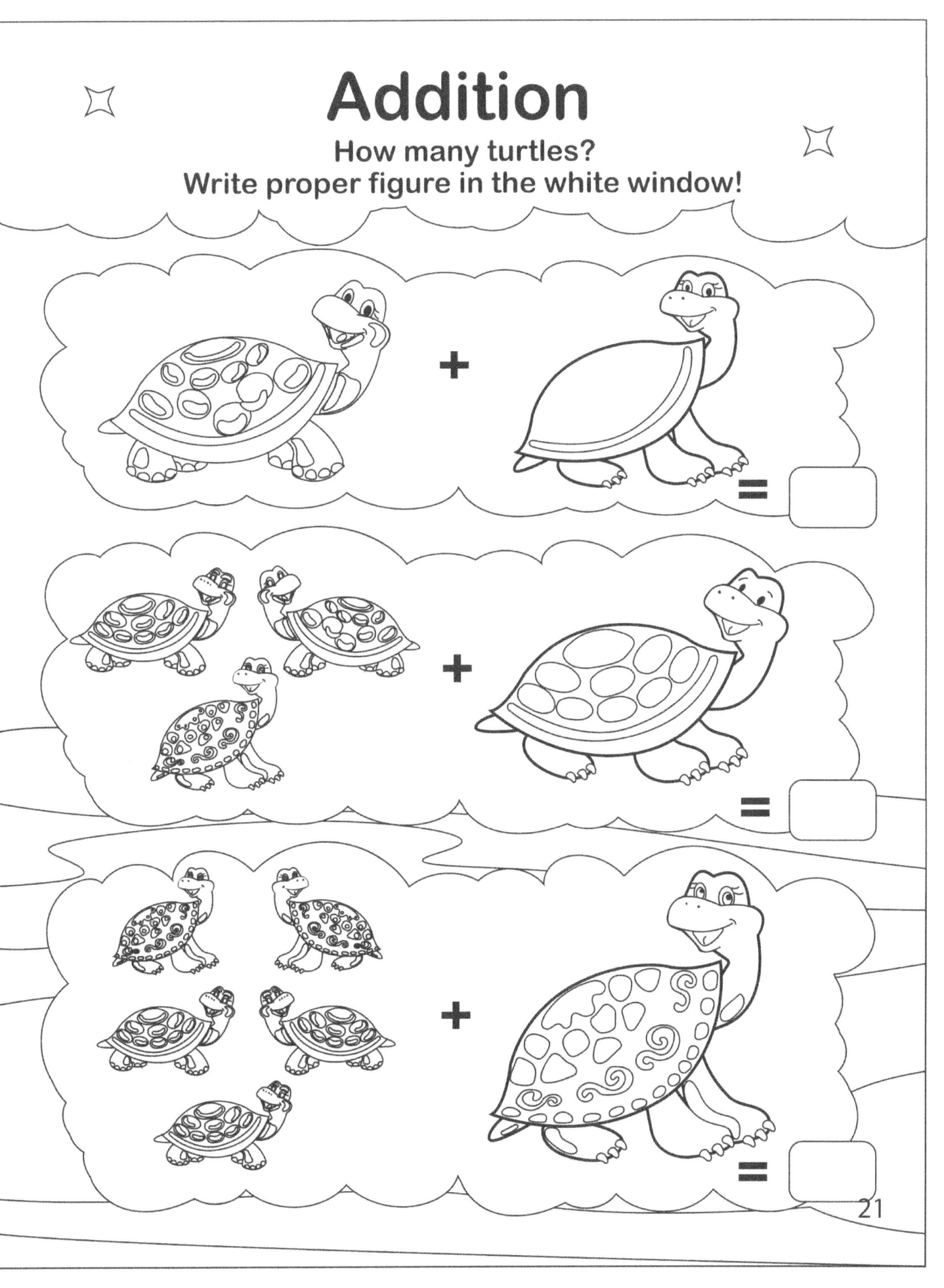

Addition

Count and circle the correct answer

5	2	3

6	4	10

8	3	4

3	1	4

22

Addition
Count and circle the correct answer

7   4   8

6   10   9

4   3   5

5   3   7

Addition
Count and circle the correct answer

10	8	7

4	9	6

2	3	6

9	8	4

24

Subtraction

How many turtles?
Write proper figure in the white window!

Subtraction
Count and circle the correct answer

| 3 | 4 | 7 |

| 6 | 3 | 8 |

| 10 | 3 | 5 |

| 2 | 1 | 4 |

26

Subtraction

Count and circle the correct answer

| 5 | 4 | 6 |

| 2 | 3 | 5 |

| 8 | 3 | 4 |

| 3 | 1 | 4 |

Subtraction

Count and circle the correct answer

4	7	3

3	4	2

8	4	5

2	1	4

Connect the numbers from 1 to 10 in correct
order by the lines and colorize the picture.

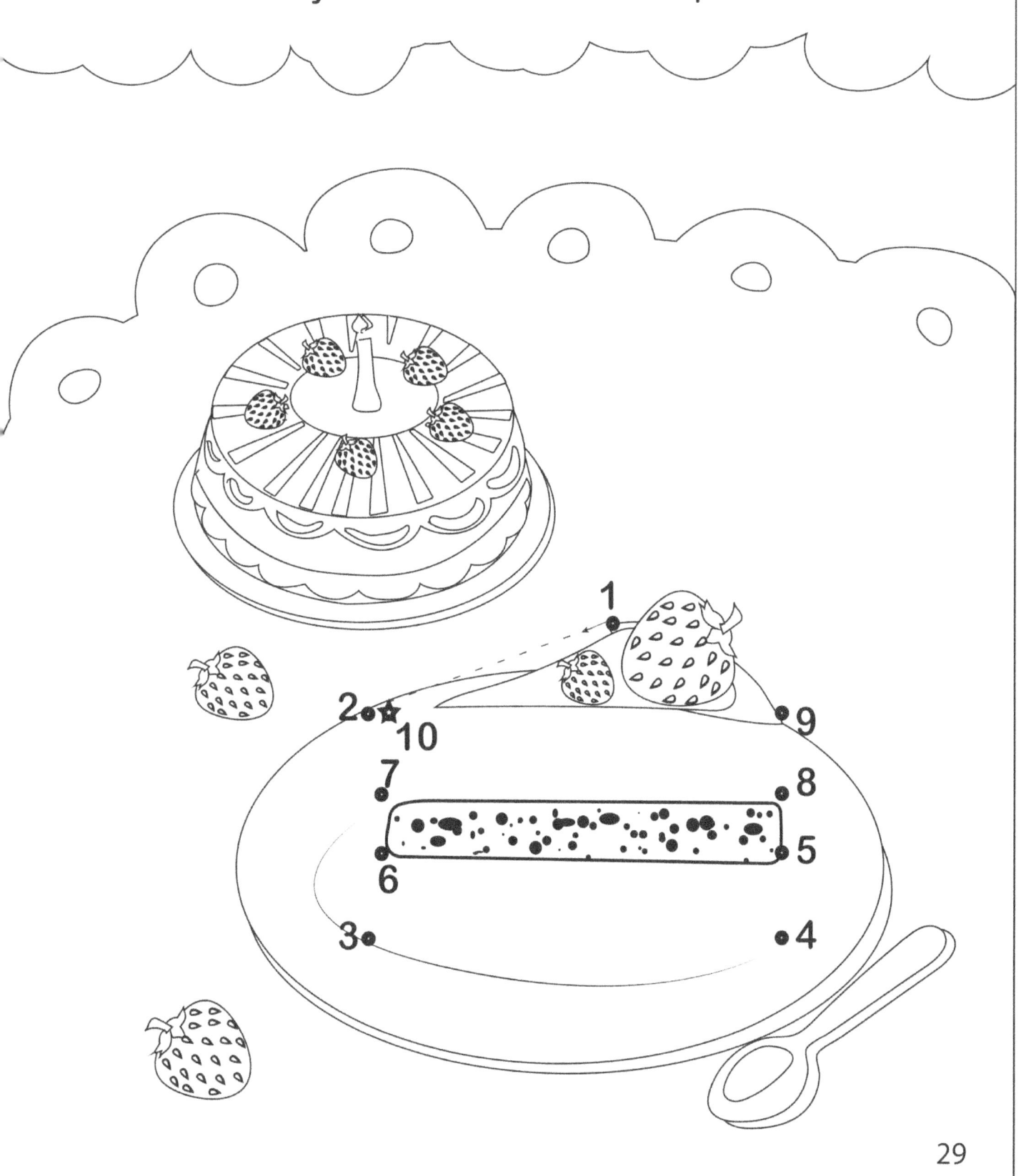

Connect the numbers from 1 to 10 in correct order by the lines and colorize the picture.

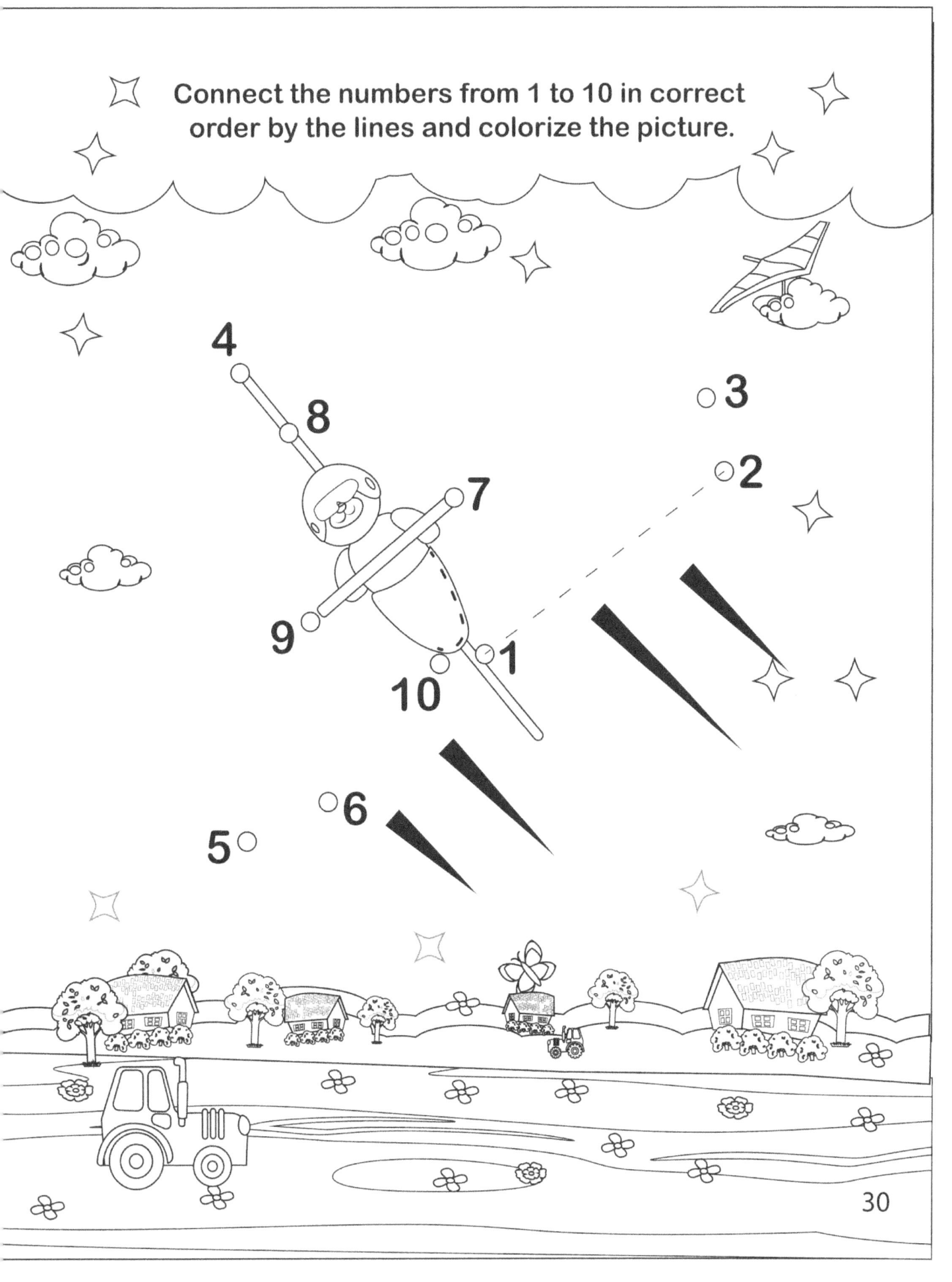

www.ingramcontent.com/pod-product-compliance
Lightning Source LLC
Chambersburg PA
CBHW080902220526
45467CB00008B/2595

* 9 7 8 1 7 1 7 3 0 7 8 8 0 *